# HIDDEN

## ANIMAL WEAPONS AND DEFENSES

by Teresa Klepinger

CAPSTONE PRESS
a capstone imprint

Published by Capstone Press, an imprint of Capstone
1710 Roe Crest Drive, North Mankato, Minnesota 56003
capstonepub.com



Library of Congress Cataloging-in-Publication Data is available on the Library of Congress website.
ISBN: 9781669078166 (hardcover)
ISBN: 9781669078111 (paperback)
ISBN: 9781669078128 (ebook PDF)

Summary: Engaging text gives readers a quick rundown of unique animals that use camouflage and mimicry as weapons and defenses.

Editorial Credits
Editor: Mandy Robbins; Designer: Dina Her; Media Researcher: Jo Miller; Production Specialist: Tori Abraham

Image Credits
Alamy: Amazon-Images, 29, David Fleetham, 25, Lee Rentz, 15, Margus Vilbas, 22, Morley Read, 5; Getty Images: Nick Dale/500px, 19, Shin Okamoto, 12; Shutterstock: Anna Veselova, 6, Artush, 7, Dan4Earth, 11, FoxGrafy, design element (throughout), Joyce LCY, Cover (top right), 17, Kris Wiktor, Cover (top left), Michal Pesata, 23, nickeverett1981, 8, 9, PACO COMO, 18, RealityImages, 26, reptiles4all, 20, Thierry Eidenweil, Cover (bottom)



Printed and bound in China. 5827

# TABLE OF CONTENTS



Words in **bold** are in the glossary.

# HIDING IN PLAIN SIGHT

A tiny frog hops from a smooth leaf to a prickly branch with moss. In minutes, its smooth skin becomes bumpy. When it jumps back to the leaf, the bumps disappear.

Animals have many ways to hide, trick, and scare. Some ways are shockingly strange!

mutable rainfrog

# NOTHING TO SEE HERE

## GIANT LEAF-TAILED GECKO

A giant leaf-tailed gecko blends in resting on a tree. But any **predator** that comes close gets a surprise.

## FUN FACT

Some people say the gecko sounds like a child when it screams.

When in danger, these geckos open their red mouths and scream. If that doesn't work, they drop their tails. Predators eat the tail, not the gecko. The tail grows back later.

## PYGMY SEAHORSE

The tiny pygmy seahorse grips **coral** with its tail. Nearby fish look for food. The seahorse isn't worried.

These colorful seahorses match the coral. Even their bumps match. They are hard to see. Scientists discovered them by surprise while studying the coral.

## FUN FACT

The pygmy seahorse comes in two color combinations: pink and purple or yellow and orange.

## GIANT SWALLOWTAIL CATERPILLAR

The giant swallowtail caterpillar looks like bird poop. But hungry spiders aren't fooled. When they come along, these caterpillars stick out a special body part. It looks like a red-orange forked tongue. It smells awful. Many spiders leave the caterpillar alone.

## LEAFY SEA DRAGON

A leafy sea dragon sways in seaweed. It changes color to blend in. Its straw-like mouth sucks up tiny shrimp.

Sea dragons look like they have seaweed growing from their bodies. Their **camouflage** is so good, they have no known predators.

# FANCY BACKPACKS

## DECORATOR CRAB

A decorator crab plucks a sponge off some coral. It sticks it to the stiff hairs on its shell. Then it grabs an **anemone** and then some seaweed. Soon, its shell is covered.

When predators swim by, they don't see decorator crabs. They see bad tasting decorations.

## FUN FACT

When a decorator crab sheds its shell, it puts the same decorations on its new shell.

## ASSASSIN BUG

The assassin bug sucks out an ant's insides. Then it glues the empty body onto its back.

Assassin bugs wear dead bodies like a backpack. This way, predators see and smell ants. They don't know an assassin bug is under the pile.

# FAKERS AND TAKERS

## DRONGO BIRD

A drongo bird sees meerkats eating tasty food. It **mimics** the meerkats' danger call. The meerkats dive into their holes.

Drongos copy many alarm calls. When the animals are scared away, the birds swoop in to steal food.

## SPIDER-TAILED HORNED VIPER

The spider-tailed horned viper lies still. It blends in with the rocks. Only the spider-like tip of its tail wags back and forth.

These vipers fool birds. The birds try to eat what they think is a spider. Instead, they get caught by the viper.

## LARGE BLUE BUTTERFLY CATERPILLAR

A large blue butterfly caterpillar makes special smells. Nearby ants think the caterpillar is an ant **larva**. They carry it to their nest.

Once the caterpillar is in an ant nest, the ants protect it. The caterpillar feasts on the real ant larvae until it turns into a butterfly.

# SHAPE-SHIFTERS

## MIMIC OCTOPUS

Mimic octopuses pretend to be other animals to defend against predators. A fish pesters the octopus. The octopus knows these fish are afraid of sea snakes. So the octopus hides in a hole. It waves two striped arms around that look like sea snakes.

## FUN FACT

The mimic octopus can look like 15 different animals.

## JUMPING SPIDER

This jumping spider doesn't want spider-eating birds to notice it. So it acts like an ant!

These spiders zigzag as they crawl. When they pause, they raise their front two legs. The legs look like ant **antennae**. The spider's body is even shaped like an ant.

## HAWK MOTH CATERPILLAR

A bird gets close to a hawk moth caterpillar. Suddenly, the bird sees a snake rise up.

When these caterpillars are frightened, their heads swell. Fake snake eyes and scales appear. It even strikes in the direction of the danger. Shocked attackers quickly move away.

# GLOSSARY

**anemone** (uh-NEM-uh-nee)—an ocean animal that looks like a plant; anemones eat other sea creatures

**antenna** (an-TEH-nuh)—feelers on an insect's head used to sense touch and smells; antennae is the word for more than one antenna

**camouflage** (KA-muh-flahzh)—a pattern or color on an animal's skin that helps it blend in with the things around it

**coral** (KOR-uhl)—a colony of tiny animals that grow together to form a structure

**larva** (LAR-vuh)—an insect at the stage of development between an egg and an adult

**mimic** (MIM-ik)—to copy or imitate

**predator** (PRED-uh-tur)—an animal that hunts other animals for food

# READ MORE

Arnold, Tedd. *Weird Animals*. New York: Scholastic Inc., 2021.

Hofer, Charles C. *Adorable but Deadly Creatures*. North Mankato, MN: Capstone Press, 2022.

Murphy, Julie. *Odd Bods: The World's Unusual Animals*. Minneapolis: Millbrook Press, 2021.

# INTERNET SITES

*It's a Bug's Life*
kids.nationalgeographic.com/videos

*San Diego Zoo Wildlife Explorers*
sdzwildlifeexplorers.org/

*Switch Zoo Animal Games*
switchzoo.com/

# INDEX



# ABOUT THE AUTHOR

Teresa Klepinger lives in the Pacific Northwest where she watches coyotes hunt, looks under rocks, and pokes things with sticks. She loves any kind of animal, teaching kids, reading books, music, and, most of all, her family.